YOUR KNOWLEDGE HAS VALUE

- We will publish your bachelor's and master's thesis, essays and papers

- Your own eBook and book - sold worldwide in all relevant shops

- Earn money with each sale

Upload your text at www.GRIN.com and publish for free

Bibliographic information published by the German National Library:

The German National Library lists this publication in the National Bibliography; detailed bibliographic data are available on the Internet at http://dnb.dnb.de .

This book is copyright material and must not be copied, reproduced, transferred, distributed, leased, licensed or publicly performed or used in any way except as specifically permitted in writing by the publishers, as allowed under the terms and conditions under which it was purchased or as strictly permitted by applicable copyright law. Any unauthorized distribution or use of this text may be a direct infringement of the author s and publisher s rights and those responsible may be liable in law accordingly.

Imprint:

Copyright © 2018 GRIN Verlag
Print and binding: Books on Demand GmbH, Norderstedt Germany
ISBN: 9783668727458

This book at GRIN:

https://www.grin.com/document/429014

Mutinda Jackson

Systems Engineering Tools and Methods

GRIN Verlag

GRIN - Your knowledge has value

Since its foundation in 1998, GRIN has specialized in publishing academic texts by students, college teachers and other academics as e-book and printed book. The website www.grin.com is an ideal platform for presenting term papers, final papers, scientific essays, dissertations and specialist books.

Visit us on the internet:

http://www.grin.com/

http://www.facebook.com/grincom

http://www.twitter.com/grin_com

Systems Engineering Analysis (Methods)

Abstract

One of the major problems linked to design and development of any multifaceted system has always been the failure of planning along with definite identification of requirements, which cause performance lack and design failure. As a result, a well-organized approach to integrated design together with the advancement of novel systems is highly required, a system referred to as systems engineering (SE). Arguably, in systems engineering, all development facets are mainly mulled over at the primary phases, not to mention that the efforts obtained are used for incessant improvement. SE may be adequately defined as the effective application of scientific and engineering efforts with the aim of transforming a functioning necessity into a clear system organization via process of need analysis, synthesis, operational analysis and allotment, design optimization, evaluation and validation. Again, this system aims at integrating related technical constraints alongside ensuring the compatibility of all physical and functional, along with programming interfaces in a way that optimizes the total definition as well as design. Subtly, SE aims at integrating reliability, safety, reliability, maintainability, serviceability, schedule, disposability to meet cost, on top of technical performance objectives.

Table of Contents

Introduction 4
Team Activity 4
The Systems Engineering Tools and Methods 4
Alternative Approaches 7
Conclusion 8
References 8

Introduction

A system refers to a set of objects that comprise of relationships between the objects (the components of the system) and their features. These features (attributes) refer to the properties of the objects alongside the casual, logical as well as random relationships that link the aforementioned components. These components (manpower, facility, software, material among others) are designed in a way that they can perform several functions, in accordance with the desired purposes and objectives of the system. Systems engineer adds value only through measuring along with reducing risk. Notably, the two main functions of systems engineering include: the verification and validation of the elements of the system so as to attain the specified quality standards or even based on a certain acceptance level. The performance of a system is mainly measured by aspects such as reliability, quality, availability, disposability and supportability among others.

Team Activity

Since complex design challenges entail facts from diverse disciplines, every individual in our group of four had to do deep research on systems engineering analysis. Our group had to coordinate during the design process on top of providing the knowledge needed in system development so as to effectively implement the design. It is widely agreed that the integrated development approach necessitates cross-functional teams that is comprised of members from all the functional areas working in proximity together and sharing details of their design portion as it progresses, on top of developing all aspects of the system simultaneously. It is a phenomenon that we had put in mind as we all knew that this would manage the overall life cycle. In order to make this task easy, each member of the team had to work on a particular method used in systems engineering; in our case we focused on reliability, maintainability, economic evaluation and serviceability.

The Systems Engineering Tools and Methods

As a technical discipline, systems engineering requires both quantitative and qualitative methods so as to understand customer needs, design robust as well as optimized systems, enable exploration of design options and even validate designs in the projected environments.

1. Reliability

Reliability as one of the main system-attributes that have great interest to system engineers, users and even logisticians can be defined as the probability of a system that performs its intended function under stated conditions, with zero failure for a particular period of time. The concept of reliability has to consist of elements such as detailed function description, time scale and the environment along with demonstrating what constitutes a failure (Girault & Valk, 2013).

Concerning design for reliability, it has been noted that system designs that are based on user needs and system design alternatives may then be formulated as well as evaluated. In this phase, reliability engineering aims at increasing system robustness via measures like, diversity, advanced diagnostics, redundancy and even modularity so as to enable rapid physical replacement (Czarnecki et al, 2000). Furthermore, there is a possibility to failure rates reduction through measures like using higher strength materials, quality components increment, moderation of the extreme environmental conditions or shortened maintenance, inspection or even overhaul intervals.

Evidently, the design analyses may include mechanical stress, radiation analyses for mechanical components, corrosion, thermal analyses for electrical and mechanical components together with EMI (electromagnetic interference) analyses or measurements for electrical sub-systems and components (Girault & Valk, 2013). One of the main obvious way to improving software reliability is through improving its quality, a context that is facilitated by more disciplined development efforts as well as test. On the other hand, reliability may also be increased through what is termed as architectural redundancy, diversity and independence. In this sense, redundancy has to be accompanied by several measures so as to ensure there is data consistency, on top of managing failure detection together with switchover.

On reliability production, most of the production issues are linked with this method, with the most significant of these being ensuring repeatability and even uniformity of production processes as well as complete unambiguous specifications for supply chain items (Buede & Miller, 2016). Other major challenges of this tool may be associated with design for manufacturability, transportation and even storage. Basically, in large software intensive systems, the information systems may be affected by issues that are related to management configuration, installation testing and that of integration.

2. Maintainability

In this context, maintainability remains to be the parameter that is concerned with the manner through which the system in use may be restored after a failure has occurred, while still considering several concepts such as preventive maintenance as well as BIT (Built-In-Test), support equipment and even maintainer skill level (Girault & Valk, 2013). Significantly, it has to be understood that while dealing with the availability requirement, the maintainability requirement has to be also invoked simply because some level of repair together with restoration to a state of mission capable has to be included.

According to diverse research, both logistics and logistic support strategies have also been noted to be closely linked, not to mention that they are independent variables at play in the requirement of availability. Accordingly, this is a phenomenon that takes the form of sparring strategies, availability of maintenance manuals, maintainer training and even identification of the basic support equipment. Since software performance has significant effects on the maintainability performance requirements, it is very vital to address software in the overall requirements. The accumulated stress mechanisms, which characterize hardware failures, do have the ability to cause a failure (Girault & Valk, 2013). On the other hand, software has been observed to exhibit behaviours that distinct operators see as a failure, thus, it is highly vital to note that users, test community, contractors and even program offices have to agree early on what constitutes a failure.

3. Economic Evaluation and Test

Significantly, testing has been defined as a mechanism that is used to assure quality of a product, system or even capability. In order to be effective, testing cannot happen only at the end phase of the development, but rather it has to be addressed incessantly through the entire life cycle. Both testing and evaluation involve the evaluation of a product right from the component level to stand-alone system level, integrated system and in some cases to system-of-system and enterprise level (Buede & Miller, 2016).

As a result, systems engineers are expected to have the ability to create strategies concerning test and evaluation to field effective and interoperable systems, which comprise of making recommendations on accreditation and certification processes (Czarnecki et al, 2000). In this context, these systems play a significant role in assisting in the development and definition of test and evaluation plans as well as procedures. Furthermore, they play a part in operational and developmental testing, observing and communicating test results, influencing re-test decisions, recommending strategies of mitigation alongside assisting customers in making system acceptance decisions.

4. **Serviceability**

Serviceability is another highly regarded attribute that has to be considered when one is designing, purchasing, manufacturing or even using a computer component or product. Serviceability can be defined as an expression of the ease with which a system, component or device may be prepared or maintained. Consequently, any early detection of the potential challenges has been seen to be one of the main issues in this respect. Research has it that, numerous systems have been noted to have the capacity of correcting challenges automatically, even prior to occurrence of the serious challenges.

In the aforementioned context, some of these problems include OSs built-in features, for instance, Microsoft Windows XP as well as the auto-protect enabled antivirus software, along with spyware detection and removal programs (Girault & Valk, 2013). Ideally, it has to be observed that maintenance of the system or device and repair operations have to be set in a manner that there will be little downtime or disruption. Serviceability is highly advantageous as it has an element or recoverability and automatic updating that keeps applications current without the intervention of the user.

Alternative Approaches

According to research, there are diverse alternatives that may be observed after evaluation that is based on their resulting change in the performance indicator. Substantially, concept and technology development may be used with an intention of exploring alternative concepts that are based on assessments of operational needs, risks, technology readiness and even affordability. Entry into such stage does not necessarily imply that DoD has committed to a novel acquisition program, but instead, it refers to the initiation of a process aimed at determining whether or not a requirement may be met at reasonable levels of technical risks as well as at affordable costs (Buede & Miller, 2016). Trade-off analyses are applied in this sense so as to examine the viable alternatives with the aim of determining the preferred approaches.

Conclusion

Suffice to say, in systems engineering, all development aspects are mainly considered at the initial phases, not to mention that the efforts obtained are employed for incessant improvement. System engineering may be adequately defined as the effective application of scientific and engineering efforts with the aim of transforming a functional requirement into a distinct system constitution via process of need analysis, synthesis, operational analysis and distribution, design optimization, evaluation and validation. Again, this system aims at integrating related technical constraints alongside ensuring the compatible nature of all physical and functional, along with programming interfaces in a way that optimizes the total definition as well as design. Subtly, system engineering aims at integrating reliability, safety, reliability, maintainability, serviceability, schedule, disposability to meet cost, on top of practical performance goals.

References

Buede, D. M., & Miller, W. D. (2016). The engineering design of systems: models and methods. New Jersey: John Wiley & Sons.

Czarnecki, K., Eisenecker, U. W., & Czarnecki, K. (2000). Generative programming: methods, tools, and applications (Vol. 16). Reading: Addison Wesley.

Girault, C., & Valk, R. (2013). Petri nets for systems engineering: a guide to modeling, verification, and applications. Springer Science & Business Media.

YOUR KNOWLEDGE HAS VALUE

- We will publish your bachelor's and
  master's thesis, essays and papers

- Your own eBook and book -
  sold worldwide in all relevant shops

- Earn money with each sale

Upload your text at www.GRIN.com
and publish for free